Total Solar Eclipse

A step-by-step guide to seeing the 2024 total eclipse and the possibilities that lie ahead.

Tyrone Ramey

TABLE OF CONTENTS

Section 1. Introduction 5

1.1 What is a solar eclipse and how does it occur?
.... 7

1.2 Why are solar eclipses rare and special? 10

1.3 What are the types of solar eclipses and how are they different? 13

Section 2. The 2024 Total Solar Eclipse...............18

2.1 When and where will it happen? 21

2.2 What will it look like and how long will it last?
.... 22

2.3 How to plan and prepare for the event.......... 25

2.4 How to safely observe and enjoy the eclipse?
.... 28

2.5 What are some interesting facts and trivia about the 2024 eclipse? 32

Section 3. Solar Eclipses in History and Culture . 37

3.1 How have solar eclipses influenced science, art, religion, and politics? 40

3.2 What are some famous or notable solar eclipses in history? 44

3.3 How do different cultures and traditions interpret and celebrate solar eclipses? 44

3.4 What are some myths and superstitions associated with solar eclipses? 51

Section 4. Solar Eclipses in the Future **55**

4.1 What are the challenges and opportunities for studying and exploring solar eclipses? 58

4.2 How will solar eclipses inspire and impact humanity in the future? 61

Section 5. Conclusion **66**

Appreciation 69

Section 1. Introduction

Imagine a time when the sun sets and a pitch-black, eerie silence falls over the sky. During the day, observers can see the stars and planets, with the moon's dark disk surrounded by a ring of fire. The animals become restless, the air colder, and people stare in amazement. This is what it's like to see a complete solar eclipse, one of the most amazing and uncommon natural events.

When the moon moves in front of the sun and casts a shadow on the earth, this phenomenon is known as a solar eclipse. The three celestial bodies' alignment determines whether an eclipse is partial, annular, or complete. The moon partially obscures the sun, creating a crescent-shaped appearance during a partial eclipse. When the moon is too far away from Earth to fully obscure the sun, an annular eclipse occurs, eclipsing the moon with a brilliant ring of light. When the moon is near enough to the earth to completely obscure the sun, a total eclipse occurs, producing a dramatic and gloomy sight.

Because solar eclipses require the exact alignment of the sun, moon, and earth, they are uncommon and unique occurrences. Since there is a 5-degree tilt in the moon's orbit around the earth, the moon is usually either above or below the plane of the earth's orbit around the sun. A solar eclipse can only occur when the moon crosses this plane at the right time and place. Variations in the moon's orbit around the earth and the earth's distance from the sun throughout the year also influence the size and form of the moon's shadow. Therefore, not every solar eclipse varies and can be observed in every region of the planet.

A complete solar eclipse will occur on Monday, April 8, 2024, visible in portions of North America, including Mexico, the US, and Canada. The moon will totally block the sun during the totality phase, which will stretch from Mazatlán, Mexico, to Newfoundland, Canada. The longest totality period will be 4 minutes and 28 seconds in Nazas, Mexico; however, it may vary depending on the location. In some parts of North America, Central America, and northern South America, there will also be partial eclipse visibility. Anyone interested in learning more about solar eclipses in general, or the complete solar eclipse of 2024, should read this guide. Topics covered will include:

➢ The history, culture, and effects of solar eclipses on human civilization and creativity
➢ The study of solar eclipses and lunar eclipses, focusing on their scientific principles and distinguishing characteristics;
➢ The future holds both exciting prospects and daunting challenges for the study and

exploration of solar eclipses, which have the potential to inspire and profoundly impact humanity.

➢ The preparation and safety tips for viewing and enjoying the 2024 total solar eclipse; and what to expect during the event.

This Guide will provide you with the knowledge and motivation you need to fully enjoy and experience the total solar eclipse of 2024, as well as solar eclipses in general, regardless of your level of interest or expertise. While we take in one of the most breathtaking and astounding views in the cosmos, we hope you will come with us on this voyage of wonder and discovery.

1.1 What is a solar eclipse and how does it occur?

An astronomical phenomenon known as a solar eclipse happens when the moon moves in front of the sun and blocks our planet with its shadow. The umbra and the penumbra are the two components of the moon's shadow. The umbra, also known as the black interior of the shadow, is where the moon completely blocks out the sun. When the moon partly obscures the sun, it creates the lighter outer region of the shadow, known as the penumbra.

You may see one of the following kinds of eclipses during a solar eclipse, depending on your location on Earth:

The moon causes a **total eclipse** when it completely blocks out the sun in the umbra's path. Because it produces a gloomy and unsettling sight in the sky, this is the most dramatic and uncommon kind of solar eclipse. The sun's corona, or outer atmosphere, is visible during a complete eclipse, even though it is normally undetectable to the unaided eye. Certain brilliant stars and planets, often obscured by the sun's brightness, become visible. Several minutes may pass during a complete eclipse, depending on your position and the moon's shadow's pace.

An **annular eclipse** occurs when you are in the umbra's path but the moon is too far away from Earth to completely obscure the sun. The "ring of fire" surrounds the moon with a brilliant ring of light. Comparable to a complete eclipse, but less spectacular and more frequent, is an annular eclipse. During an annular eclipse, the sun is still too bright for you to see the stars and planets, much less the sun's corona. Similar to a complete eclipse, an annular eclipse lasts for a few minutes.

When you are in the penumbra, or path, where the moon is partly blocking the sun, you experience a **partial eclipse**. As a result, the sun takes on a crescent shape in the sky, which varies in size and form as the moon passes over it. The most frequent and least impressive kind of solar eclipse is a partial one, as it doesn't significantly alter the environment's temperature or light. During a partial eclipse, the duration can vary depending on your location and the extent of the moon's coverage.

The occurrence of a solar eclipse requires the moon to be positioned between the sun and the earth during a new moon. The moon's orbit around the planet is tilted by approximately 5 degrees; thus, not every new moon results in a solar eclipse. Instead, the moon is usually placed above or below the plane of the earth's orbit around the sun. A solar eclipse can only occur when the moon crosses this plane at the right time and place. The term "node" refers to this junction point, and the moon's orbit has two nodes: the ascending node and the descending node. If one of these nodes is in close proximity to the new moon, a solar eclipse may happen.

Additionally, because of the moon's somewhat eccentric orbit, its distance from Earth changes year-round. The term "apogee" refers to the moon's furthest position from Earth, while "perigee" denotes its closest point. The moon's distance from Earth determines the size and form of its shadow, and consequently, the type of solar eclipse that occurs. A total eclipse occurs when the moon is near perigee, when it appears larger and may fully block the sun. An annular eclipse occurs when the moon is close to its apogee, appears smaller, and is unable to completely hide the sun. During a solar eclipse, the sun, moon, and earth align precisely, creating an intricate and captivating natural event. In the next section, we shall discuss why solar eclipses are unique and uncommon, as well as how often they happen.

1.2 Why are solar eclipses rare and special?

One of the most breathtaking natural events that anyone may see is a solar eclipse. They take place when the moon directly faces the sun, causing a shadow to fall on Earth. Solar eclipses vary in their characteristics. Some are partial, in which the moon merely covers a portion of the sun. Others are annular, which means that a ring of light surrounds the black disk because the Moon is too far away from Earth to completely encircle the Sun. There are also total eclipses, in which the moon totally obscures the sun's disk, resulting in a short moment of darkness and the appearance of the sun's thin outer atmosphere, known as the corona.

Total solar eclipses are the rarest and most dramatic of all. They only take place when the moon is precisely positioned in its orbit and at the proper distance from Earth to the sun. This necessitates a careful balancing act between a number of variables, including the inclination of the Earth's axis, the size, shape, and tilt of the lunar orbit, and the seasonal fluctuations in the Earth-Sun distance. Due to these reasons, total solar eclipses only happen once every eighteen months on average anywhere on Earth, but they are only viewable from a particular spot once every 375 years on average. Furthermore, the period of totality—that is, the amount of time during which the Moon totally covers the Sun—is quite brief, lasting anywhere from a few seconds to a maximum of almost seven minutes.

Millions of people in North America will have the unique chance to view this cosmic beauty during the unusually complete solar eclipse of 2024. It will be the first total solar eclipse to be observed from Mexico and Canada since 1991 and the first to traverse the entire United States from coast to coast since 1918. The Moon will totally obscure the Sun in the route of totality, which is a narrow strip that stretches from the Pacific to the Atlantic Ocean and crosses over 12 states, numerous major cities, and portions of Mexico and Canada. The total path is about 15,000 kilometers (9,300 miles) long. The longest totality will occur in Mazatlán, Mexico, and will last for 4 minutes and 28 seconds; however, the exact length may vary depending on the location. Parts of Europe, Africa, South America, Central America, and North America will also be able to see the eclipse in partial form.

Solar eclipses are not only gorgeous and uncommon, but they are also important for science. They provide a rare opportunity for scientists and astronomers to examine the sun's corona, which is often obscured by the brightness of the star. The outermost layer of the sun's atmosphere, known as the corona, is where the solar wind originates, and temperatures may reach millions of degrees. Flares and mass ejections in the corona can alter the Earth's magnetic field and trigger geomagnetic storms. Scientists may get additional insight into the composition, activity, and impact of the sun on the solar system by studying the corona during a complete solar eclipse.

Sunscreen is necessary to shield your eyes from the sun's damaging rays while watching a solar eclipse. Even in cases of partial eclipses, you should never gaze directly at the sun without wearing protective eyewear. During the short moment of totality, when the moon totally covers the sun, this is the only time you may stare at the sun without protection. But you must take care to turn away or reapply your protection as soon as the sun emerges. Using specialized eclipse glasses or filters, which block out the majority of the sun's light and let you see the moon's silhouette, is the ideal method to see a solar eclipse. Another option is to use a pinhole projector, which is a simple gadget that displays the picture of the sun on a wall or screen. Using a piece of paper, a pin, and a cardboard box, you can create your own pinhole projector. Additionally, a live web feed of the eclipse is available from a number of websites, including timeanddate.com and NASA.

You shouldn't miss this unique and uncommon event— a solar eclipse. It's an opportunity to take in the splendor and awe of nature while also learning more about the sun and how it affects the earth. For many individuals in North America, seeing the complete solar eclipse in 2024 will be a once-in-a-lifetime chance to see this incredible phenomenon. Plan beforehand, choose a comfortable spot to watch, and take in the show!

1.3 What are the types of solar eclipses and how are they different?

When the moon moves in front of the sun and blocks some or all of the sun's light from reaching Earth, it's called a solar eclipse. Solar eclipses vary in their characteristics. There are three basic varieties of solar eclipses: total, partial, and annular, depending on the Sun, Moon, and Earth's relative positions and distances from one another. Every variety has its own distinct qualities and origins.

Total solar eclipse: The most spectacular and uncommon kind of solar eclipse occurs when the Moon totally obscures the Sun's disk, resulting in a short period of darkness and the revelation of the Sun's thin outer atmosphere, known as the corona. Only when the Sun, Moon, and Earth are perfectly aligned and the Moon is at or near its closest point to Earth (perigee) can there be a complete solar eclipse. This necessitates the exact coincidence of a number of variables, including the inclination of the Earth's axis, the seasonal fluctuations in the Earth-Sun distance, and the size, shape, and tilt of the lunar orbit. Due to these reasons, total solar eclipses only happen once every eighteen months on average anywhere on Earth, but they are only viewable from a particular spot once every 375 years on average.

Furthermore, the period of totality—that is, the amount of time during which the Moon totally covers the Sun—is quite brief, lasting anywhere from a few seconds to a maximum of almost seven minutes. The sky becomes black during totality, giving the impression that it is dawning or dusk, and the planets and stars may be seen. The temperature decreases, and plants and animals could act as if night has fallen. The moon's black disk is surrounded by a luminous ring of colorful or white light, which is the corona, which is often obscured by the sun's brilliance. The outermost layer of the sun's atmosphere, known as the corona, is where the solar wind originates, and temperatures may reach millions of degrees. Solar flares and coronal mass ejections, which have the ability to alter the Earth's magnetic field and trigger geomagnetic storms, are also produced in the corona. Scientists may get additional insight into the composition, activity, and impact of the sun on the solar system by studying the corona during a complete solar eclipse. You must shield your eyes from the sun's damaging rays in order to see a complete solar eclipse safely. Even in cases of partial eclipses, you should never gaze directly at the sun without wearing protective eyewear. During the short moment of totality, when the moon totally covers the sun, this is the only time you may stare at the sun without protection. But you must take care to turn away or reapply your protection as soon as the sun emerges.

Using specialized eclipse glasses or filters, which block out the majority of the sun's light and let you see the moon's shadow, is the ideal method to see a complete solar eclipse. Another option is to use a pinhole projector, which is a simple gadget that displays the picture of the sun on a wall or screen. Using a piece of paper, a pin, and a cardboard box, you can create your own pinhole projector.

Additionally, a live web feed of the eclipse is available from a number of websites, including [timeanddate.com] and [NASA Live].

Partial solar eclipse: The most frequent kind of solar eclipse is a partial one, in which the moon just partly obscures the sun's disk, giving it a crescent form. When the Sun, Moon, and Earth are not exactly aligned, but the Moon still casts a shadow on Earth, this phenomenon is known as a partial solar eclipse. The relative locations of the sun and moon in the sky determine the shadow's size and form. Observers can view a partial solar eclipse from a wider region compared to a total solar eclipse. The sun's corona remains visible, and the sky does not become darker. Even still, the sun continues to appear as a brilliant crescent, making it hazardous and blinding to stare at without protection. To properly see a partial solar eclipse, you also need eclipse glasses, filters, pinhole projectors, or live streams—the same tools required for a complete solar eclipse observation. Even when the sun is partly obscured, you should never gaze directly at it since this might result in severe eye damage.

Annular solar eclipse: An annular eclipse is a unique type of solar eclipse that occurs when the Moon obscures the Sun's core, leaving the Sun's outside borders in the shape of a "ring of fire," or annulus. An annular solar eclipse occurs when the Moon aligns perfectly with the Sun, Moon, and Earth at its furthest point (apogee). The moon seems smaller and does not entirely hide the sun because of its greater distance from Earth. This causes the moon to appear as a dark disk above a bigger, brighter disk, giving the appearance of a ring around the moon. An annular solar eclipse can be observed from a specific path on Earth, with the annulus lasting up to 12 minutes. The sun's corona is not visible, and the sky does not get black. If you gaze at the ring of fire without protection, it may still cause harm to your eyes due to its intense brightness. Use eclipse glasses, filters, pinhole projectors, or live broadcasts, just like you would for a complete or partial solar eclipse, to safely see an annular solar eclipse. Even during an annular eclipse, you should never gaze directly at the sun since this might result in irreversible eye damage.

Section 2. The 2024 Total Solar Eclipse

You won't want to miss the incredible spectacle that is the complete solar eclipse in 2024. It will be the first total solar eclipse to be observed from Mexico and Canada since 1991 and the first to traverse the entire United States from coast to coast since 1918. The eclipse will occur on Monday, April 8, 2024, lasting around four hours and twenty-six minutes. The Moon will totally obscure the Sun in the route of totality, which is a narrow strip that stretches from the Pacific to the Atlantic Ocean and crosses over 12 states, numerous major cities, and portions of Mexico and Canada. The total path is about 15,000 kilometers (9,300 miles) long. The length of totality, or the moment when the Moon fully covers the Sun, varies according to the location; in Mazatlán, Mexico, it will last for the greatest period of time, 4 minutes and 28 seconds. Parts of Europe, Africa, South America, Central America, and North America will also be able to see the eclipse in partial form.

A complete solar eclipse is a unique and exceptional chance to take in the majesty and beauty of the natural world while also learning more about the sun and how it affects the earth. The sky becomes black during a complete solar eclipse, giving the appearance of dawn or dusk and revealing the planets and stars. The temperature decreases, and plants and animals could act as if night has fallen. The moon's black disk is surrounded by a luminous ring of colorful or white light, which is the corona, which is often obscured by the sun's brilliance. The outermost layer of the sun's atmosphere, known as the corona, is where the solar wind originates, and temperatures may reach millions of degrees. The corona also produces solar flares and coronal mass ejections, which can alter the Earth's magnetic field and trigger geomagnetic storms. Scientists may get additional insight into the composition, activity, and impact of the sun on the solar system by studying the corona during a complete solar eclipse.

You must shield your eyes from the sun's damaging rays in order to see a complete solar eclipse safely. Even in cases of partial eclipses, you should never gaze directly at the sun without wearing protective eyewear. During the short moment of totality, when the moon totally covers the sun, this is the only time you may stare at the sun without protection. But you must take care to turn away or reapply your protection as soon as the sun emerges. Using specialized eclipse glasses or filters, which block out the majority of the sun's light and let you see the moon's shadow, is the ideal method to see a complete solar eclipse. Another option is to use a pinhole projector, which is a simple gadget that displays the picture of the sun on a wall or screen.

Using a piece of paper, a pin, and a cardboard box, you can create your own pinhole projector. Additionally, a live web feed of the eclipse is available from a number of websites, including [timeanddate.com] and [NASA Live].

You must prepare ahead of time and choose a decent viewing spot if you want to get the most out of your eclipse experience. The route of totality will traverse some of North America's most picturesque and significant locations, such as the Grand Canyon, the Rocky Mountains, the Mississippi River, and Niagara Falls. You may even go to several of the cities—Dallas, Indianapolis, Cleveland, Montreal, and Quebec City—that will see totality. Millions of people will go to these locations to see the eclipse, so you should be prepared for extreme crowding and activity. Make sure to reserve your lodging and transportation well in advance, and be ready for delays, cancellations, and traffic jams. Checking the weather forecast and having a backup plan in case of clouds or rain are also essential. The optimal locations to see the eclipse, together with the precise timings and lengths of each eclipse phase, may be found using this interactive map.

You shouldn't miss this unique and uncommon event—a complete solar eclipse. It's an opportunity to take in the splendor and awe of nature while also learning more about the sun and how it affects the earth. For many individuals in North America, seeing the complete solar eclipse in 2024 will be a once-in-a-lifetime chance to see this incredible phenomenon. Plan beforehand, choose a comfortable spot to watch, and take in the show!

2.1 When and where will it happen?

On **Monday, April 8, 2024**, there will be a complete solar eclipse. It will last around **4 hours and 26 minutes.** The eclipse will occur in the Pacific Ocean, south of Alaska, at **10:42 a.m. PDT (17:42 UTC),** and in the Atlantic Ocean, east of Newfoundland, at **3:08 p.m. EDT (19:08 UTC).** The Moon will totally obscure the Sun in the route of totality, which is a narrow strip that stretches from the Pacific to the Atlantic Ocean and crosses over 12 states, numerous major cities, and portions of Mexico and Canada. The total path is about 15,000 kilometers (9,300 miles) long. The length of totality, or the moment when the Moon fully covers the Sun, varies according to the location; in Mazatlán, Mexico, it will last for the greatest period of time, 4 minutes and 28 seconds. Parts of Europe, Africa, South America, Central America, and North America will also be able to see the eclipse in partial form.

2.2 What will it look like and how long will it last?

Skywatchers throughout North America will never forget 2024, when they see a complete solar eclipse, a celestial sight of breathtaking grandeur. A complete eclipse is an incredibly life-changing event, in contrast to a partial eclipse, in which the moon just partially obscures the sun. Let's explore the specifics of what to anticipate during this once-in-a-lifetime occasion, with an emphasis on the eclipse's length and visual features.

A transcendental metamorphosis

For those fortunate enough to be in the line of totality on April 8, 2024, anticipation will surely grow closer to that day. The moon will seem to take a little bite off of the edge of the sun as the eclipse begins slowly. This first stage, called first contact, is the beginning of an amazing dance between the stars.

The moon will gradually block out the sun's bright surface with each passing instant. Enchanting twilight will cover the countryside as the sunshine fades. It's possible for the temperature to gently decrease and for animals to behave strangely in reaction to the shifting light.

Baily's beads are a phenomenon that becomes noticeable when totality approaches. These are brief but brilliant bursts of sunshine that are peeping through valleys on the craggy edge of the moon. This serves as a reminder of the sun's enormous power, which is obscured for a short while.

The Absolute Moment

When the eclipse reaches totality, it will reach its peak. The day will have entered an unsettling twilight as the moon will have totally obscured the sun's face. There will be an unusual, even bizarre, light falling over the countryside. Stars that are ordinarily invisible during the day may briefly come into view.

The dim outer atmosphere of the sun, known as the corona, will soon become visible. The eclipsed sun is surrounded by this ethereal ring of pearly white light, which highlights the sun's enormous structure. In addition, prominences—flaming gas loops that shoot out of the sun's surface—might show out against the night sky, giving even more drama to the celestial show.

An important aspect of the eclipse experience is the length of totality, or the period of time when the sun is totally hidden. During the 2024 eclipse, some areas may experience a remarkable 4 minutes and 28 seconds of maximum totality, bringing a hearty serving of darkness. This gives you plenty of time to take in the grandeur of the occasion and get amazing pictures. On the other hand, the length of totality will vary depending on where you are on the route. Areas nearer the path's centerline will have longer totality times, while areas further away will have shorter totality times.

The final phase of the eclipse

The cosmic dance reaches its last act as the moon starts to travel away from the sun. The prominences and corona vanish from sight, and Baily's beads come back into view in a stunning show that heralds the sun's slow return. The conclusion of totality and the return of daylight are marked by the second contact, when the sun appears from behind the edge of the moon.

The stages of partial eclipses continue, with the moon gradually exposing more and more of the sun until it is once again completely visible. The world goes back to normal as the landscape takes on its accustomed hues.

Witness a truly remarkable experience during the complete solar eclipse in 2024. The chance to see the moon completely obliterate the sun and the stunning vistas of the prominences and corona will make this an unforgettable experience. Mark your calendars, carefully consider where to observe, and be ready to be in awe of the cosmic wonder of the complete 2024 solar eclipse.

2.3 How to plan and prepare for the event

Experts expect the Great North American Eclipse, also known as the total solar eclipse of 2024, to be a remarkable astronomical event. It is a once-in-a-lifetime event to see the sun fully disappear behind the moon, transforming the day into a dreamy twilight. But cautious planning and preparation are essential if you want to enjoy this celestial phenomenon to the fullest. This guide will provide you with the information and instructions you need to guarantee a secure and breathtaking eclipse experience.

Step 1: ascertain if you are on the path of totality.

For seeing a solar eclipse, the path of totality is critical. The only region in North America where the Moon will totally block the Sun on April 8, 2024, is this small strip that runs from Mexico to Canada. Online tools abound to verify whether your region is eligible for a total eclipse. You may find out where you are and if it's in the path by using websites like the NASA Eclipse webpage and interactive maps from trustworthy astronomy groups.

Step 2: Choose Your Viewing Location

The weather can have a big impact on your eclipse experience, even if you're in the path of totality. You might get a feel for what to expect by looking up the area's historical weather patterns for April. Ideally, you want to find a spot where you can set up for the event comfortably, away from any structures or trees, and with a clear view of the western horizon. Public parks, wide open spaces, or designated viewing zones often provide excellent viewing locations for eclipses.

Step 3: Assemble the Necessary Materials

Prepare the required equipment for a pleasant eclipse viewing experience. The following list will help you make sure you're ready:

> ➢ ***Secure Sun Eclipse Goggles:*** Never, not even for a little instant, look straight at the sun. Ensure solar eclipse glasses are ISO-rated and adhere to global safety regulations. With the protection of these glasses, you may view the many stages of the eclipse without risk.
> ➢ ***Cozy seats or blanket:*** To unwind and savor the experience, make sure you have cozy seats or a blanket because the eclipse sequence can run for several minutes.
> ➢ ***Sun Protection:*** The sun's powerful UV rays might still be present throughout the brief totality phase. For further protection, bring sunglasses, a hat, and sunscreen.

> ➤ **_Snacks and Water:_** Bring enough food and drink to last the entire viewing session.
> ➤ **_Camera (Optional):_** For best results, investigate suitable camera settings and filters if you intend to take pictures of the eclipse. Recall that the requirement for safe solar eclipse glasses for direct viewing does not disappear while using a camera.
> ➤ **_Extra Power Pack:_** Make sure you have enough power for all of your electronics, especially your phone, which you use to take pictures and view maps.

Step 4: Familiarize yourself with the Eclipse Schedule.

There are different phases to a complete solar eclipse, and each one presents a different kind of sky show. Comprehending these stages will enhance your viewing experience:

> ➤ **_Partial Eclipse:_** As the moon begins to eat away at the sun's disc, the eclipse quietly starts. This first stage may take more than an hour.
> ➤ **_Totality:_** The ultimate achievement of an eclipse is when the moon fully envelops the sun. The duration of this period can vary depending on your location, from a few seconds to many minutes. The sky becomes noticeably darker

during totality, revealing the sun's corona—a thin, hazy coating of plasma.

- ➢ **Partial Eclipse Resumes:** When the Moon moves forward, it returns to the partial eclipse phase, progressively revealing the Sun. This marks the end of totality.
- ➢ **Eclipse End:** When the Moon fully separates from the Sun's disc, the eclipse comes to an end, and the sky returns to normal during the day.

Step 5: Take Charge of Your Observations

Observing eclipses frequently draws sizable groups. It's crucial to show respect for other onlookers. To guarantee an excellent vantage point and prevent obstructing others' views, arrive early. To save the environment, dispose of any rubbish in an appropriate manner.

You can make sure that your experience seeing the 2024 total solar eclipse is safe, unforgettable, and breathtaking by adhering to these guidelines and adopting a respectful and well-prepared mindset.

2.4 How to safely observe and enjoy the eclipse

The Great North American Eclipse, often known as the complete solar eclipse of 2024, promises to be a remarkable astronomical event. Seeing the sun entirely disappear and be replaced by the corona's ethereal brightness is a once-in-a-lifetime sight. Safe observation is crucial, however. Here is a thorough guide to help you appropriately appreciate this cosmic wonder:

Keeping Your Eyes Safe: A Top Priority

Even during a partial eclipse, staring straight at the sun may result in lifelong eye damage. The strong radiation from the sun may burn the retina, resulting in blindness or loss of eyesight. With the exception of totality, when the moon totally blocks the sun's direct light, this is accurate for every stage of the eclipse.

The golden rule is this: **Never gaze directly at the sun or through any unprotected optical instrument, such as telescopes, binoculars, or sunglasses.**

Using specially made solar filters is the only safe technique to see the sun directly during the partial phases. These filters considerably lower sun radiation to acceptable levels while still meeting international safety requirements (ISO 12312-2).

Selecting Appropriate Solar Filters

> ➤ *Reputable Vendors:* Get your solar filters from suppliers that you can trust to uphold

safety regulations. Verify credentials with organizations such as the American Astronomical Society (AAS).

➢ **Inspection:** Carefully check your filters for damage, pinholes, and scratches before usage. A defect of any size may jeopardize safety. Throw away any broken filters.

➢ **Verify** that the filters are properly labeled to indicate that they are intended for direct viewing of the sun. Seek out the certification for ISO 12312-2.

Safe-Viewing Techniques

➢ **Sun Filter Glasses:** Depending on the required degree of light reduction, these specifically made glasses are available in a variety of tints. Select eyewear designed especially for watching solar eclipses.

➢ **Shade #14 Welder's Glass:** Use care while using Shade #14 Welder's Glass for safe sun observation. Verify that the glasses are genuine #14, and double-check the shade rating. Seasoned eclipse watchers should use caution due to potential differences in filter quality.

➢ **Indirect viewing techniques:** Take into account indirect viewing techniques as a secure and interesting substitute. Through these techniques, you may see the eclipse without having to stare directly at the sun. They project the picture of the sun onto a surface. Here are two well-liked choices:

- Pinhole Projector: Using cardboard and aluminum foil, make a basic pinhole projector. To see how the eclipse is progressing, project the picture of the sun onto a white surface.
- Solar Projection Telescope: A solar projection telescope is a more advanced choice. A detailed view of the eclipse may be obtained using this equipment, which displays a magnified picture of the sun on a screen.

Making the most of your Eclipse experience

Once you've figured out how to watch safely, be ready to be amazed! To improve your eclipse experience, consider the following advice:

- ➢ **Select your site:** Congratulations if you're on the path of totality! Get ready to see the magnificent corona and the sun's total disappearance. A partial eclipse will still occur if you're not in the path. For detailed eclipse maps, use internet resources (eclipse route 2024).
- ➢ **Get There Early:** Congestion may build up at viewing points, especially along the line of totality. To guarantee an excellent viewing position, arrive early.

- ➢ ***Weather Takeaways***: Even if you have no control over the weather, be prepared for unforeseen circumstances. Just in case, bring a light jacket, a hat, and sunscreen.
- ➢ ***Seize the moment:*** With the right filters, photographers may take breathtaking pictures of the eclipse. Do some advanced research on suitable filters and approaches.
- ➢ ***Savor the experience:*** An eclipse is an uncommon occurrence in space. Take some time to marvel at the wonders of the cosmos and share the experience with loved ones.

You'll be well on your way to safely seeing the complete 2024 solar eclipse and making lifelong memories if you heed these safety precautions and preparatory advice. Recall that making appropriate observations guarantees you will be able to see the eclipse and tell others about the marvel for many years to come.

2.5 What are some interesting facts and trivia about the 2024 eclipse?

As the much-awaited solar eclipse of 2024 draws near, let's lift the veil on the sky and discover fascinating details that give this astronomical event more depth. Fasten your seatbelts, fellow astronomers, and join us as we set off on an uncomplicated cosmic adventure.

North America's Great Eclipse:

Skywatchers from Mexico to Canada will be able to see the solar eclipse of 2024, named The Great North American Eclipse, when its path of totality sweeps over North America. Why? Because skywatchers from Mexico to Canada will be able to see it when its path of totality sweeps over North America.

Imagine a 2,500-mile ribbon of darkness, or a cosmic highway, extending from the Pacific Northwest to the Atlantic Northeast. Cosmopolitan cities like Buffalo, Cleveland, Dallas, and Indianapolis are in the limelight.

The Eclipse's Return:

2017 was the final complete solar eclipse that the contiguous United States could see. Seven years later, the moon's shadow is once again visible in our sky. On **April 8, 2024**, the Sun, Moon, and Earth will align in a cosmic pas de trois. Mark your calendars for that day.

The Celestial Sphere Dance:

The moon's shadow travels over the surface of the earth at an astounding **1,500 miles per hour** during totality. If you blink, you may miss it!

Think of the moon as a cosmic ninja who is subtly obstructing the sun's light. What was the outcome? This is an amazing demonstration of cosmic choreography.

The Search for Eclipse Chasers:

Eclipse chasers travel across continents in pursuit of the shadow cast by the moon. From the Arctic Circle to the Sahara Desert, they have seen eclipses. Why? Each eclipse is unique in its own way. The corona reveals different mysteries, and each landscape reacts uniquely.

The Impact of the Diamond Ring: An Astral Request

The diamond ring effect, a dazzling point of light, appears just before totality. It's like a proposal from the cosmos to us.

The last rays of the sun cast a celestial engagement ring as they passed across lunar valleys.

Accept wonderment!

Eclipses in Mythology and History:

Prehistoric societies saw eclipses as signs of a heavenly struggle between light and dark.

According to Norse mythology, eclipses were caused by wolves chasing the sun and moon. The cosmic wolf, Fenrir, was the main character.

The **Mystery of the Eclipse** Be attentive to the ground as totality approaches. Observers might catch sight of unseen **shadow bands**—wave-like patterns of light and dark.

They remain a topic of discussion among scientists. Is there turbulence in the atmosphere? Or a ripple in the cosmos?

Safety of Eclipses: Defend Those Eyes!

Safety comes first, and we cannot emphasize this enough. When there is an eclipse, avoid looking straight at the sun.

Make a simple pinhole projector or use authorized **eclipse glasses**. The cosmos should respect your eyes.

The soundtrack for Eclipse:

During totality, pay careful attention. The birds stop talking, and the crickets start their nighttime concert. As if in awe of the cosmic display, the Earth itself seemed to be holding its breath.

The universe reminds us that we are stardust. Remind yourself that we are a part of the cosmic fabric as the moon's shadow moves over the landscape. Watching the motion of the heavenly bodies, we are on a faint blue dot. Formerly, far-off stars formed our atoms.

As April 8, 2024 approaches, inform your friends and family about this cosmic story. assemble in open spaces, on roofs, or on slopes. Allow the eclipse to work its magic, serving as a momentary reminder that we are all just dust particles dancing in a vast cosmic dance.

Section 3. Solar Eclipses in History and Culture

Humanity has been enthralled with the breathtaking sight of a complete solar eclipse for millennia, during which the Moon briefly envelops the Sun. This section delves into the complex cultural importance and historical background of solar eclipses, highlighting the ways in which these astronomical occurrences have influenced our perceptions of the cosmos and our role in it.

Previous Documents and Forecasts:
Ancient China is where some of the first solar eclipse observations were made and recorded. These incidents are documented in records as early as 2137 BCE; some of them only say, "The Sun has been eaten." These findings were essential in the creation of precise calendars, which were a necessary tool in agricultural cultures. The Chinese also held the belief that the emperor's health and well-being were related to eclipses. There is a legend that two astronomers, Hsi and Ho, were put to death for not seeing a solar eclipse, which goes to show how important these celestial events are.

Mythology and omens:

All around the world, societies have created their own tales and lore to account for the striking reduction in daylight that occurs during a solar eclipse. For example, in ancient Greece, a solar eclipse was seen as a divine rebuke, whereas the Mayans connected it to the Sun being devoured by a heavenly jaguar. Similar to this, the eclipse is linked to the demon Rahu in Hindu mythology. Rahu is killed while attempting to take the nectar of immortality, and his head is constantly following the Sun. These tales capture the wonder and terror evoked by eclipses, which are often seen as signs of imminent catastrophe or divine intervention.

Scientific Tipping Points:

Scientific progress has also been greatly aided by solar eclipses. The ancient Babylonians developed their knowledge of lunar cycles via the observation of eclipses. A solar eclipse in the sixth century BCE demonstrated how these occurrences might even have an impact on historical events by mediating a conflict between the Lydians and the Medes. Later, Greek astronomer Hipparchus made a crucial discovery that advanced our knowledge of the solar system: he calculated the distance between Earth and the Moon using a solar eclipse.

Contemporary Scientific Findings:

Powerful telescopes have made it possible to examine the Sun's corona, the outermost layer of its atmosphere that is only seen during a complete solar eclipse, in greater detail. The study of solar activity and its effects on Earth was greatly aided by these observations. Furthermore, measurements performed during a solar eclipse in 1919 established the validity of the theory of relativity, one of the pillars of contemporary physics. One of the main predictions of Einstein's ground-breaking theory—that light bends around enormous objects like the Sun—was confirmed by this eclipse.

Prospects:

Solar eclipses maintain their scientific and cultural importance. They excite the public's astonishment and interest while giving scientists the chance to study the Sun and the corona. A remarkable occurrence, the complete solar eclipse of 2024 is expected to be visible across North America. It will provide insight into the dynamic dynamics of our solar system and serve as a reminder of the intricate web that connects humans and the universe.

This section gives the 2024 solar eclipse its historical and cultural background, emphasizing both its ongoing curiosity and the significant influence it has had on our comprehension of the cosmos.

3.1 How have solar eclipses influenced science, art, religion, and politics

April 8, 2024, is approaching, and it looks to be a stunning astronomical event—a complete solar eclipse. For millennia, this cosmic ballet, where the Moon momentarily obscures the Sun, has enthralled people. However, solar eclipses have considerably more effects than just briefly making the day seem shorter. These historical occurrences have had a significant impact on politics, science, art, and even religion.

Solar Eclipses and Science: An Inquiry into Understanding

For millennia, solar eclipses fueled scientific exploration by offering a look into the unknown. Prehistoric societies employed them to measure time and forecast heavenly movements. For example, the Babylonians created a complex lunar calendar because they painstakingly documented eclipses for generations.

Scientists pursued a better understanding of eclipses to advance astronomy. The nature of the sun and moon was a topic of discussion between Aristotle and Thales of Miletus, among other ancient Greek philosophers. Eclipses provided important support for their celestial theories.

The discovery of telescopes transformed the study of eclipses. A solar eclipse in 1859 allowed scientists to examine the sun's chromosphere, a hitherto unseen component of its atmosphere. This discovery led to a better understanding of solar activity and its effects on Earth.

Solar eclipses are still useful scientific phenomena today. They are used by astronomers to investigate the sun's corona, a weak outer layer that is often hidden by the intense light of the sun. Eclipses also reveal dim planets and stars usually hidden by the Sun's brilliance. We can better comprehend our role in the cosmos and the origin of our solar system because of the information obtained from these observations.

Eclipses of the Sun and Art: A Source of Inspiration

For millennia, the breathtaking spectacle of a complete solar eclipse has inspired artists. These astronomical occurrences have inspired creative expression across civilizations, from prehistoric cave drawings showing celestial phenomena to Renaissance engravings catching the exact intricacies of an eclipse.

Many civilizations viewed eclipses as signs of ill luck or the wrath of God. Eclipse-themed artwork, which often shows heavenly conflicts or hideous monsters eating the sun, reflects these beliefs.

Modern art follows a distinct method. Photographers record the corona's amazing splendor, while artists capture the dramatic change from day to night in totality with vivid hues. Sculptors have even created works that resemble the ethereal appearance of the corona.

Undoubtedly, the eclipse of 2024 will spark a new generation of creative works. The eclipse will surely make its impact on the art world, whether it's a thought-provoking sculpture reflecting the sun's momentary absence or a stunning snapshot of the diamond ring effect—the first glimmer of sunshine peeping through the moon's shadow.

Religion and Solar Eclipses: Creating Myths and Beliefs

Throughout history, solar eclipses have been deeply significant to religions. Many civilizations viewed them as conflicts between heavenly bodies, signs of impending change, or otherworldly occurrences. Eclipses were thought to be produced by a heavenly monster consuming the sun in ancient China. To frighten the dragon away, they would smash cymbals and pound drums. In certain Native American tribes, eclipses were associated with conflicts between the Sun and the Moon, leading to sacrifices being offered. Eclipses have also influenced religious calendars. Because they could foresee eclipses, civilizations were able to schedule religious rituals and celebrations around them.

Eclipses still have an impact on religion today. While some religious communities see them as signals or cautions, others regard them as chances for introspection. Religious organizations that connect astronomical occurrences to prophecy or divine intervention may find special meaning in the 2024 eclipse.

Politics and Solar Eclipses: A Weapon for Prestige and Power

Some emperors in antiquity believed they could control eclipses and used this ability to establish their dominion. Some emperors in antiquity believed they could control eclipses and used this ability to establish their dominion. Correctly predicting an eclipse may increase a leader's reputation for discernment and divine favor.

Military strategy has also taken advantage of eclipses. Certain conflicts, for instance, are scheduled to take place during eclipses to gain an edge over the opponent.

In the present period, eclipses no longer have as much influence on politics. Nevertheless, these occasions have the power to thrill and captivate people. The 2024 eclipse is likely to be a big deal in the media, drawing a lot of attention and possibly increasing travel to places within its line of totality.

The 2024 total solar eclipse is expected to profoundly affect politics, art, society, and religion. Experiencing this celestial show up close offers an opportunity to interact with the universe's incredible power.

3.2 What are some famous or notable solar eclipses in history?

Solar eclipses are those celestial shows in which the sun is briefly obscured by the moon; they have fascinated people for thousands of years. These celestial occurrences, which range from prehistoric omens to scientific discoveries, have profoundly impacted our collective awareness. Let's examine some of the most amazing solar eclipses in recorded history, when it seemed as if the skies were pausing to reveal hidden meanings.

1. **August 21, 2017's Great American Total Solar Eclipse**

    > ➢ *Path of Totality:* From Oregon to South Carolina, a 70-mile shadow stretched across the country.

> - **Limited Accessibility:** Millions of Americans were able to see this complete eclipse, in contrast to most that occur in isolated areas.
> - **Historical Significance:** This was the first complete solar eclipse to occur in the United States since 1776.
> - **Safety Tip:** Whenever possible, use eye protection during an eclipse, with the exception of the short totality when the moon completely blocks the sun.

2. The Ugarit Eclipse of Ancient Times (1375 B.C.)

> - **Clay Tablet Chronicles:** written on a prehistoric clay tablet discovered in Ugarit, which is now in Syria.
> - **The Sun Put to Shame:** During this complete eclipse, Mesopotamian historians said that the sun was "put to shame."
> - **Disputed Date:** March 5, 1223 B.C., or May 3, 1375 B.C.? Researchers are still trying to solve the enigma.

3. The Assyrian Eclipse (763 B.C.)

- ➤ *Empire in Darkness:* The sun disappeared for a whole five minutes in ancient Assyria, which is today Iraq.
- ➤ *Cosmic Drama:* Can you imagine the shock of seeing the sun suddenly vanish?
- ➤ *Historical Context:* An important juncture in Assyrian history coincided with the eclipse.

4. The eclipse at Eddington, May 29, 1919

- ➤ British astronomer Arthur Eddington used this eclipse to confirm Einstein's **general relativity theory**.
- ➤ *Bending Light:* Einstein's predictions were verified as starlight was bent by the sun's gravity.
- ➤ *Scientific Triumph:* A historic eclipse that fundamentally altered our perception of the universe.

5. The Cairn L Petroglyphs of the Lough Crew (3340 B.C.)

- ➤ Spiral petroglyphs found in Ireland suggest the oldest known solar eclipse.
- ➤ These prehistoric painters might have interpreted the eclipse as a celestial dance.

> *Timeless Connection:* People have been staring at the same heavenly delights for millennia.

6. The music from Eclipse

> *Quiet Birds:* Aware of the cosmic upheaval, birds become quiet during totality.
> *Chirping of the crickets:* As dusk falls, crickets start their nightly concert.
> *Earth Holds Its Breath:* As if to acknowledge the cosmic display, the planet itself pauses.

7.Cosmic Reminders

> *Stardust Connection:* Our atoms formed in far-off stars, and we reside on a faint blue dot.
> *The Dance of Worlds:* We see the vast cosmic ballroom as the moon's shadow moves over Earth.

Recall that every eclipse inspires us to gaze up, establish a connection with the universe, and be in awe of our position in the enormous fabric of existence—whether we are novice astronomers or avid observers.

3.3 How do different cultures and traditions interpret and celebrate solar eclipses?

Humanity has been enthralled by the breathtaking sight of a total solar eclipse for millennia, during which the moon completely blocks out the sun. These cosmic occurrences have woven themselves into the cultural tapestry of communities all around the globe, inspiring a variety of interpretations and lively traditions, going beyond the scientific surprise. Let's examine how different societies have historically interpreted and observed solar eclipses.

1. **Old Astronomical Theories:**

    ➢ *Mesoamerica:* The Maya society in Mesoamerica viewed eclipses as contests between heavenly bodies. A solar eclipse represented a conflict between the goddesses of the moon, Ix Chel, and the sun, Kinich Ahau. According to them, offerings and ceremonies could placate the gods and bring the universe back into harmony.

    ➢ *China:* In the past, Chinese astronomers believed that eclipses were a message from the sky disapproving of the emperor's activities. Special rites arranged maintained the empire's prosperity and placated the cosmic order.

> ***Ancient Greece:*** The Greeks ascribed lunar eclipses to the deeds of legendary animals. Some said the sun was consumed by a gigantic snake, while others reasoned that it happened because Helios, the god of the sun, lost control of his chariot.

2. Omens and Predictions:

> Many societies in Africa, India, and Europe viewed eclipses as omens of ill luck or imminent tragedy. Eclipses may cause disease, starvation, or the passing of a monarch. People performed certain rites or acts to ward off any bad luck.

> Astrology was also important. Some civilizations thought that the way the stars lined up during an eclipse might affect what happened in the future. Based on the Sun's and Moon's positions in the zodiac, astrologers may interpret the eclipse and forecast possible consequences.

3. Modern Celebrations:

➢ Modern science has superseded several of the previous theories on eclipses. Still, there is curiosity about these occurrences.

➢ A lot of people are interested in solar eclipses and travel to the path of totality to see this amazing phenomenon. Researchers use solar eclipses to study the Sun's corona, a faint layer of plasma that is typically hidden by the Sun's bright light. These eclipses are often the subject of festivals and educational activities that foster a feeling of community and shared astonishment.

4. Building a Cultural Bridge:

With its path over North America, the total solar eclipse of 2024 offers a unique chance to commemorate the variety of cultures that surround these astronomical occurrences. Local groups may organize events to showcase the many traditions and interpretations of eclipses. Educational programs may draw attention to the rich cultural legacy and scientific relevance of these cosmic events.

We may better appreciate the lasting impact of solar eclipses if we understand the many ways that societies have seen and reacted to them.

In addition to providing an incredible opportunity to see an amazing cosmic show, the eclipse in 2024 will honor the universal human sense of wonder and curiosity about the universe.

3.4 What are some myths and superstitions associated with solar eclipses?

One of the most breathtaking natural events that anyone may see is a solar eclipse. They take place when the moon directly faces the sun, causing a shadow to fall on Earth. But not everyone is aware of the scientific rationale for this cosmic occurrence. Many societies have created their own stories and superstitions throughout history to explain the abrupt departure of the sun. Certain beliefs serve as a source of direction, protection, or prophecy; others are motivated by fear, curiosity, or veneration. In this section, we shall look at some of the most fascinating and varied myths and superstitions from throughout the globe.

Eating the Sun: The concept that a heavenly being or a demon is attempting to consume the sun is one of the most prevalent motifs in myths about solar eclipses, leading to the eclipse. For instance, many in Vietnam thought that the sun had been swallowed by a gigantic frog. Two wolves called Skoll and Hati hunted the Sun and the Moon across the sky, occasionally catching and devouring them. People would beat pots and drums to frighten away a dragon that was said to attack the sun during an eclipse in ancient China. Rahu, an eternal demon, attempted to take the gods' elixir of life but the gods slew him before he could consume it. The Sun and the Moon would sometimes consume his severed neck and return after his head shot upwards.

Gods Enraged: Another prevalent motif in stories about solar eclipses is the notion that the sun conceals or withholds its light as a symbol of fury or retribution because it is angry or dissatisfied with humanity. For instance, in ancient Greece, people thought that a solar eclipse portended catastrophes and devastation because it was a sign of the gods' wrath. Several Native American tribes, such as the Tewa and the Pomo, viewed the Sun as a powerful god who would depart the sky to travel to his home in the underworld when displeased or in need of rest. Certain African societies, like the Batammaliba of Togo and Benin, believed that the Sun and the Moon were a married couple and that a solar eclipse indicated a conflict between them. The populace would endeavor to settle their personal disputes and promote harmony between the sun and the moon.

Bad Omens: Throughout history, solar eclipses have been associated with various negative omens such as starvation, war, illness, and death. For instance, in ancient Mesopotamia, people would designate a stand-in king to shoulder the blame and shield the legitimate monarch in the event of a solar eclipse, which they saw as a harbinger of the monarch's approaching death. People in certain regions of India fasted and took baths before and after the eclipse in order to cleanse themselves because they thought that the eclipse would taint food and water. People would wear metal amulets or red ribbons to ward off the evil impact of solar eclipses, believing them to bring ill luck or damage to pregnant women and their unborn children in several European nations, such as France and Italy.

Good Luck: Fortunately, not all beliefs and superstitions about solar eclipses are unfavorable. Certain societies viewed solar eclipses as a sign of opportunity, blessing, or good fortune. For instance, during a solar eclipse, people in certain regions of Italy used to think that flowers planted there would bloom brighter and more vividly than normal. A solar eclipse was seen as a moment of rebirth and renewal by various Native American tribes, including the Lakota and the Navajo, and as such, they would hold rites and rituals to deepen their spiritual ties to the sun. Many Asian nations, including Thailand and Japan, believed that making a wish during a solar eclipse increased the likelihood of it coming true.

For millennia, solar eclipses have captivated and motivated people, inspiring a plethora of varied and intricate stories and superstitions. Even if some of these ideas appear bizarre or illogical to us in the modern day, the cultural and historical environment in which they were formed is reflected in them. They also demonstrate how people have attempted to understand the meaning and importance of solar eclipses, as well as deal with the awe and anxiety associated with seeing them. We may enjoy the scientific and artistic grandeur of this natural phenomenon, as well as the cultural and spiritual heritage of the legends around it, as we get ready to witness the complete solar eclipse of 2024.

Section 4. Solar Eclipses in the Future

Don't miss the breathtaking experience of the total solar eclipse in 2024. It will be the first total solar eclipse to be observed from Mexico and Canada since 1991 and the first to traverse the entire United States from coast to coast since 1918. It won't be the last, however. Solar eclipses are a common but infrequent natural occurrence on Earth. There are three varieties of solar eclipses: total, partial, and annular, depending on the Sun, Moon, and Earth's relative positions and distances from one another. Every variety has its own distinct qualities and origins. This section will examine some of the solar eclipses that will occur in the next fifty years, along with their locations and dates.

In 2026, an annular and complete solar eclipse will occur. An annular solar eclipse occurs when the Moon aligns perfectly with the Sun, Moon, and Earth at its furthest point (apogee). The moon seems smaller and does not entirely hide the sun because of its greater distance from Earth. This causes the moon to appear as a dark disk above a bigger, brighter disk, giving the appearance of a ring around the moon. On February 17, 2026, an annular solar eclipse will occur, marking the first eclipse of 2026. Observers in Antarctica, Africa, South America, the Pacific, Atlantic, and Indian oceans, as well as in Antarctica itself, can witness a partial eclipse. The second solar eclipse of 2026 is scheduled for August 12, 2026, and it will be total. While a partial eclipse may be seen from Europe, Africa, North America, the Atlantic, Arctic, and Pacific oceans, it can also be seen from Greenland, Iceland, Spain, Russia, and a small portion of Portugal.

In 2027, an annular and complete solar eclipse will occur. On February 6, 2027, there will be an annular solar eclipse, the first of the year 2027. A partial eclipse may be seen from South America, Africa, Antarctica, the Pacific, Atlantic, and Indian oceans, as well as from Chile, Argentina, the Atlantic, Indian, and Antarctic oceans. On August 2, 2027, a complete solar eclipse will occur, marking the second of the year 2027. With a totality length of up to 6 minutes and 23 seconds, it will be among the longest and most beautiful total solar eclipses in recorded history. A partial eclipse can be seen in Europe, Africa, Asia, Australia, the Pacific Ocean, the Indian Ocean, and the Atlantic Ocean. It can be seen in Morocco, Spain, Algeria, Tunisia, Libya, Egypt, Saudi Arabia, Yemen, Somalia, Ethiopia, Kenya, Tanzania, Mozambique, Madagascar, and a small portion of Australia.

In 2028, an annular and complete solar eclipse will occur. On January 26, 2028, there will be an annular solar eclipse, the first of the year 2028. Observers in South America, Antarctica, the Pacific, Atlantic, and

Indian oceans, as well as in Ecuador, Peru, Brazil, Suriname, French Guiana, and Antarctica, may witness a partial eclipse. On July 22, 2028, there will be a complete solar eclipse, the second of the year 2028. Australia and New Zealand will be able to see it, and the Pacific and Indian oceans, as well as Antarctica, will be able to see a partial eclipse.

In 2030, there will be a total of two solar eclipses. On November 25, 2030, the first solar eclipse of 2030 will take place. A partial eclipse may be seen from Africa, Antarctica, Australia, the Pacific Ocean, and the Indian Ocean. It can be seen in Botswana, South Africa, Lesotho, Eswatini, Mozambique, Madagascar, Australia, and New Zealand. On December 31, 2030, the second solar eclipse of 2030 will take place. A partial eclipse may be seen in Europe, Africa, Asia, Australia, the Pacific Ocean, and the Indian Ocean. It can be seen in Algeria, Libya, Egypt, Sudan, Ethiopia, Somalia, Yemen, Saudi Arabia, Oman, Iran, Pakistan, India, China, and North Korea.

Only a few solar eclipses are expected to occur within the next fifty years. There will be several more, each with unique characteristics and places.

4.1 What are the challenges and opportunities for studying and exploring solar eclipses?

Skywatchers across North America can anticipate a celestial show when the Great North American Eclipse arrives in 2024. The upcoming solar eclipse on April 8th will mark the final opportunity to witness a complete solar eclipse from the continental United States until 2044. This book offers a thorough examination of this astronomical phenomenon, explaining its scientific foundation, providing useful advice for safe viewing, and highlighting the opportunity it gives for new scientific discoveries.

Possibilities and Difficulties for Solar Eclipse Research

Although researching total solar eclipses offers a unique set of possibilities and problems for scholars, it also provides an amazing view into the workings of the cosmos.

Problems:

Transient Events: The moment of totality, when the moon fully obscures the sun's dazzling face, only lasts a few minutes. Given this limited time, researchers are under tremendous pressure to collect as much data as they can in the constrained amount of time given this little window.

Logistical Obstacles: The path of totality, or the small band over which a complete eclipse appears, frequently passes through isolated or poorly populated areas. In these kinds of places, it may be difficult and costly to set up and run sophisticated equipment.

Weather Dependence: During an eclipse, cloud cover may seriously impede observations. Even a small layer of clouds can obscure important features, leading to an unsuccessful mission.

Possibilities include:

> ➤ *Discovering the Sun's Secrets:* During totality, the corona—the sun's thin outer atmosphere—becomes visible. Research into the corona enables understanding of solar wind, flares, and coronal mass ejections, all of which may influence Earth's magnetosphere and communication systems.
> ➤ *Uncovering the Secrets of the Universe:* During a complete eclipse, there is a short window of darkness that makes it possible to see stars and objects close to the sun that are normally obscured by its brightness. This makes it possible for astronomers to examine these objects in great detail, which may result in fresh

insights about the genesis and development of our solar system and other systems.

> *Testing General Relativity Theories:* According to Einstein's theory of gravity, big objects such as the sun have the ability to bend light. Starlight traveling close to the sun is somewhat diverted during a complete eclipse. With remarkable accuracy, scientists may test the general relativity predictions by detecting this deflection.

Technological Developments:

Thankfully, new developments in technology are assisting researchers in overcoming these obstacles and taking advantage of the potential that solar eclipses bring. The following are some significant developments:

> *Remote Observing Networks:* Instruments such as telescopes may be remotely controlled after being positioned strategically along the path of totality. This lessens the logistical load and permits the collection of more data.

> *High-Speed Data Acquisition:* These days, equipment is capable of gathering enormous volumes of data quickly. This makes it possible for

scientists to collect comprehensive data during the brief totality periods.

> ***Better Image Processing Techniques:*** By removing artifacts and noise from eclipse observations, sophisticated software enables researchers to extract even more information from the data they have gathered.

Researchers and skywatchers alike have a fantastic chance. We may gain a greater knowledge of our sun, its impact on our planet, and the vast cosmos that surrounds us by recognizing the possibilities and challenges it presents.

4.2 How will solar eclipses inspire and impact humanity in the future?

One of the most breathtaking natural events that anyone may see is a solar eclipse. They take place when the moon directly faces the sun, causing a shadow to fall on Earth. But not everyone is aware of the scientific rationale for this cosmic occurrence. Many societies have created their own stories and superstitions throughout history to explain the abrupt departure of the sun. Certain beliefs serve as a source of direction, protection, or prophecy; others are motivated by fear, curiosity, or veneration. In this subsection, we will look at how solar eclipses will affect science, culture, and spirituality in the future and how they will inspire mankind.

Science: Researchers have a rare chance to examine the sun and its impact on Earth during solar eclipses. The decrease in solar energy that occurs during a solar eclipse causes changes in the ionosphere, which is the uppermost layer of the atmosphere that reflects radio waves and includes charged particles. Systems used for communication and navigation, such as GPS and shortwave radio, may be impacted by these modifications.

Scientists may get additional insight into the structure, dynamics, and interactions of the ionosphere with the solar wind—the stream of charged particles that emanates from the sun—by examining it during a solar eclipse. Scientists may also examine the sun's corona, which is the outermost layer of its atmosphere and is often obscured by the brightness of the sun, during solar eclipses. The corona reaches temperatures of millions of degrees Celsius, serving as the source of solar flares and coronal mass ejections. These occurrences have the ability to alter the Earth's magnetic field and trigger geomagnetic storms, which may interfere with satellites, electrical systems, and auroras. Scientists may get additional insight into the sun's activity, composition, and effects on the solar system by studying the corona during a solar eclipse. Scientists may test new tools and methods, such as spectrographs, polarimeters, and high-resolution cameras, to learn more about the sun and its atmosphere during solar eclipses. Additionally, solar eclipses allow scientists to carry out studies that require complete darkness, such as discovering exoplanets or measuring the gravitational deflection of light.

Culture: For millennia, solar eclipses have captivated and motivated people, inspiring a wide range of stories and superstitions. While some of these ideas have been superseded or altered by reason and scientific understanding, others are still prevalent and significant in various regions of the globe. Nevertheless, since they pique our interest and inspire our imagination, solar eclipses continue to have a cultural and creative influence on people. Many literary and creative works, including poetry, books, paintings, music, and movies, have been inspired by solar eclipses. The poems "Darkness" by Lord Byron, "A Connecticut Yankee in King Arthur's Court" by Mark Twain, "Impression, Sunrise" by Claude Monet, "Eclipse" by Pink Floyd, and "2001: A Space Odyssey" by Stanley Kubrick are a few examples. Additionally, solar eclipses have been used in topics and genres including science fiction, fantasy, horror, and romance as metaphors, emblems, and motifs. Shakespeare's

"King Lear," Bram Stoker's "Dracula," J.R.R. Tolkien's "The Lord of the Rings," and Stephenie Meyer's "Twilight" are a few examples. Numerous festivals, events, and customs have also been influenced by solar eclipses. Examples of these include the Eclipse Festival in Australia, the Eclipse Party in Chile, and the Eclipse Prayer in Indonesia.

Spirituality: Because solar eclipses inspire awe, wonder, and a feeling of connectedness to the divine, they have also been considered spiritual and religious occurrences. Some view solar eclipses as opportunities for transformation and growth, divine signals, or cosmic messages. Solar eclipses may also be used by some as opportunities for introspection, prayer, or meditation because they help them feel more connected to the universe, the natural world, and themselves. Additionally, some people believe that solar eclipses have astrological and numerological significance and have an impact on one's personality, destiny, and interpersonal connections. In addition, some individuals engage in rites, ceremonies, or practices—such as fasting, providing sacrifices, or chanting—that are associated with their beliefs, customs, or traditions.

Solar eclipses are a common but infrequent natural occurrence on Earth. They provide an opportunity to take in the majesty and beauty of nature while also learning more about the sun and how it affects our world. They also have a significant influence on science, culture, and spirituality within mankind. We may enjoy the scientific and artistic grandeur of this natural phenomenon, as well as the cultural and spiritual heritage of the legends around it, as we get ready to witness the complete solar eclipse of 2024.

Section 5. Conclusion

You won't want to miss the amazing spectacle that will be the 2024 complete solar eclipse. It will be the first total solar eclipse to be observed from Mexico and Canada since 1991 and the first to traverse the entire United States from coast to coast since 1918. It won't be the last, however. Solar eclipses are a common but infrequent natural occurrence on Earth. There are three varieties of solar eclipses: total, partial, and annular, depending on the Sun, Moon, and Earth's relative positions and distances from one another. Every variety has its own distinct qualities and origins.

This guide covers the following subjects:

➢ **Describe a solar eclipse and explain how it occurs.** We've now covered the mathematical and scientific foundations of solar eclipses, as well as how the Sun, Moon, and Earth's alignment and distance from one another affect their occurrence. Additionally, we have discussed the many stages of a solar eclipse and how they alter how the sun and the sky seem.

➢ **Why solar eclipses are unique and uncommon:** We have looked at the factors that make solar eclipses uncommon occurrences that differ in terms of frequency, length, and location. Additionally, we've spoken about the historical and cultural importance of solar eclipses and how they've influenced humankind's advancements in science, the arts, and spirituality.

➢ **We have provided information on the date, location, and duration of the 2024 total solar eclipse.** Additionally, we've included some pointers and recommendations on how to locate the finest viewing spots and resources, as well as how to properly watch and enjoy the eclipse.

➢ **Future solar eclipses:** We've looked at some of the eclipses that will occur over the next 50 years, as well as their locations and timing. We have also highlighted some of the characteristics and locations of these eclipses, as well as the ways in which they will influence and motivate future generations of humans.

Solar eclipses are a common but infrequent natural occurrence on Earth. They provide an opportunity to take in the majesty and beauty of nature while also learning more about the sun and how it affects our world. They also have a significant influence on science, culture, and spirituality within mankind. We may enjoy the scientific and artistic grandeur of this natural phenomenon, as well as the cultural and spiritual heritage of the legends around it, as we get ready to witness the complete solar eclipse of 2024. I really hope that this tutorial has improved your understanding of and enjoyment of solar eclipses and that you will come see this incredible phenomenon with us.

Appreciation

Thank you for reading this guide on the 2024 solar eclipse. I hope that you have found it informative, interesting, and helpful.

I appreciate your interest and curiosity in this natural phenomenon, and I hope that you will join us in witnessing this amazing event.

Solar eclipses are rare and special occasions that offer us a chance to marvel at the beauty and wonder of nature, as well as to learn more about the Sun and its effects on our planet. They also have a profound impact on humanity, in terms of science, culture, and spirituality. As we prepare to watch the 2024 total solar eclipse, we can appreciate the scientific and artistic beauty of this natural phenomenon, as well as the cultural and spiritual legacy of the stories that surround it.

I would love to hear your feedback on this guide, and how I can improve it for future readers. Please leave me a 5-star review and share your thoughts and suggestions. Your opinion matters to me, and I value your support and encouragement.

Thank you again for choosing this guide, and I hope that you will enjoy the 2024 total solar eclipse.

Remember to stay safe, find a good viewing location, and have fun!